ALL ABOUT BABY SEA LIONS

by Martha E. H. Rustad

PEBBLE
a capstone imprint

Pebble Emerge is published by Pebble, an imprint of Capstone.
1710 Roe Crest Drive
North Mankato, Minnesota 56003
www.capstonepub.com

Library of Congress Cataloging-in-Publication Data
Names: Rustad, Martha E. H. (Martha Elizabeth Hillman), 1975- author.
Title: All about baby sea lions / by Martha E.H. Rustad.
Description: North Mankato, Minnesota : Pebble, [2022] | Series: Oh baby! | Includes bibliographical references and index. | Audience: Ages 5-8 | Audience: Grades K-1 | Summary: "There's a new baby joining the colony. It's a sea lion pup! Learn all about baby sea lions, including what they eat, what they weigh, how they're raised, and how big they grow"—Provided by publisher.
Identifiers: LCCN 2021002632 | ISBN 9781663908032 (hardcover) | ISBN 9781663908001 (pdf) | ISBN 9781663908025 (kindle edition)
Subjects: LCSH: Sea lions—Infancy—Juvenile literature.
Classification: LCC QL737.P63 R87 2022 | DDC 599.79/751392—dc23
LC record available at https://lccn.loc.gov/2021002632

Image Credits
BluePlanetArchive.com: Mark Conlin, 13; Shutterstock: Andrea Izzotti, 12, Anna Om, 16, Blue Ice, 6, Daniel Avram, 19, Eric Isselee, back cover, Foto 4440, cover, 7, 9, imageBROKER.com, 14, kongsak sumano, 5, Leonardo Gonzalez, 11, Monkey Business Images, 20, NATTHAWAT PHROMTHAISONG, 21, Sydni Josowitz, 17, wildestanimal, 15

Editorial Credits
Editor: Alison Deering; Designer: Jennifer Bergstrom; Media Researcher: Tracy Cummins; Production Specialist: Tori Abraham

Table of Contents

Words in **bold** are in the glossary.

A NEWBORN PUP

A new baby is here! It is a sea lion **pup**. The baby cuddles close to its mother. It grew inside her for about a year. It drinks milk from her body.

Sea lions are ocean animals. They come on land to have babies. Family groups gather on beaches or on ice. A family is made up of one male and many females.

The mother stays close to her baby for several days. She leaves when it's time to eat. She must find food in the ocean.

Pups gather in groups while their mothers are hunting. They walk on their **flippers**. They play and make noise.

When a mother comes back, she calls for her baby. The pup hears her. It makes its own sound. The mother and pup know each other's noises.

Pups weigh 13 to 20 pounds (6 to 9 kilograms) when they are born. They are born with baby fur. It keeps them warm until they grow **blubber**. Blubber is a layer of fat.

Pups shed their baby fur after a few months. Adult fur grows in. Adult males have **manes** of thick fur, just like lions. That is how sea lions got their name.

9

LEARNING AND GROWING

Pups have to learn a lot in their first year. They learn by copying adults. Mothers teach their babies to swim. They start in shallow water. Then they learn to swim in deep water.

Sea lions can dive deep underwater. Some stay under as long as 30 minutes. They close their **nostrils**. They use their flippers to swim.

Sea lions can swim very fast. Their top speed is 25 miles (40 kilometers) per hour. Swimming fast helps them catch **prey**.

Sea lions also swim fast to get away from **predators**. Sharks and whales hunt and eat sea lions. Young sea lions learn to escape. They swim fast and quickly get out of the water.

Pups eat and grow a lot. At about
2 months old, they start to eat fish.
They drink milk for at least six months.

Pups learn how to hunt. They eat fish, squid, crabs, and octopus. They have sharp teeth for grabbing food. They use their **whiskers** to sense where food is.

Tired sea lions pile together to rest. They haul their heavy bodies out of the water onto land or ice. They sometimes gather and float together in the water.

Sea lions are noisy animals. They roar, bark, honk, and squeak. They can be heard from miles away. They also make noise underwater!

ALL GROWN UP

Sea lions are fully grown after one year. They live for 20 to 30 years. Some stay with their family group. Others form new groups.

When sea lions have babies, they return to the beaches where they were born. They have their own babies there.

SEA LIONS AND FISH

Sea lions swim fast to catch fish in the ocean. Pretend to be a sea lion with your friends as the fish. How many fish can you catch?

What You Need

- an open area for playing
- cones or other items to make boundaries
- a group of players

What You Do

1. Use the cones to mark out the "ocean." This can be a square or rectangular area.

2. Choose one player to be the sea lion. This person stands in the middle of the "ocean."

3. All the other players are fish. Have them stand at one end of the ocean.

4. When the sea lion says, "Swim, little fish!" all the fish try to make it to the other side of the ocean. If they get caught, they become sea lions. If they make it across, they get to try again.

5. Keep playing until all the fish are caught. Then choose a new sea lion.

Glossary

blubber (BLUH-buhr)—a thick layer of fat under the skin of some animals that keeps them warm

mane (MAYN)—long, thick hair that grows on the head and neck of some animals

nostril (NOSS-truhl)—an opening in the nose used to breathe and smell

predator (PRED-uh-tur)—an animal that hunts other animals for food

prey (PRAY)—an animal hunted by another animal for food

pup (PUHP)—a young animal

whisker (WISS-kur)—a long, stiff hair growing on the face and bodies of some animals

Read More

Adamson, Heather. *Sea Lions*. Minneapolis: Bellwether Media Inc., 2018.

Gish, Ashley. *Sea Lions*. Mankato, MN.: Creative Education, 2019.

Orr, Tamra B. *Seal or Sea Lion?* Ann Arbor, MI: Cherry Lake Publishing, 2020.

Internet Sites

Dolphin Research Center: Sea Lion Facts for Kids
dolphins.org/kids_sea_lion_facts

National Geographic Kids: California Sea Lions
kids.nationalgeographic.com/animals/mammals/
california-sea-lion/

San Diego Zoo Animals & Plants: Sea Lion
animals.sandiegozoo.org/animals/sea-lion

Index